Mathematical Olympi

for

Elementary School 1

My First Book of Mathematical Olympiads – *First Grade*

(Workbook Plus)

Educational Collection *Magna-Scientia*

My First Book of Mathematical Olympiads

Mathematical Olympiads *for* Elementary School

1

First Grade

(Workbook Plus)

Michael Angel C. G., Editor

Preface

The Mathematical Olympiads for the First Grade of Elementary School discussed here are none other than the *Mathematical Olympiads for Schoolchildren "Unikum"*, which are held every year in the city of Lipetsk since 2010, and organized by the Faculty of Physics, Mathematics and Computer Science of Lipetsk State Pedagogical University and the Center for Continuing Education of Children "Strategy". Likewise, these Olympiads consist of two rounds, a qualifying round and a final round, both consisting of a written exam. The problems included in this book correspond to the final round of these Olympiads.

The present edition called Workbook Plus seeks to consolidate the mathematical skills acquired with the previous workbook since it includes new variants of problems as well as more challenging tasks. In this workbook has been compiled all the Olympiads held during the years 2011-2020 and is especially aimed at schoolchildren between 6 and 7 years old, with the aim that the students interested either in preparing for a math competition or simply in practicing entertaining problems to improve their math skills, challenge themselves to solve these interesting problems; or it could even be used for a self-evaluation in this competition, trying the student to solve the greatest number of problems in each exam in a maximum time of 1 hour 10 minutes. It can also be useful for teachers, parents, and math study circles. The book has been carefully crafted so that the student can work on the same book without the need for additional sheets, what will allow the student to have an orderly record of the problems already solved.

Each exam includes a set of 8 problems from different school math topics. To be able to face these problems successfully, no greater knowledge is required than that covered in the school curriculum; however, many of these problems require an ingenious approach to be tackled successfully. Students are encouraged to keep trying to solve each problem as a personal challenge, as many times as necessary; and to parents who continue to support their children in their disciplined preparation. Once an answer is obtained, it can be checked against the answers given at the end of the book.

Sincerely,

The editor

Contents

Problems

Olympiad 2011

(II Mathematical Olympiad "Unikum")

Problem 1. Which of the following numbers is the greatest: 2010, 2110, 2100, 2001 or 2011?

Problem 2. Which of these shapes has the greatest area?

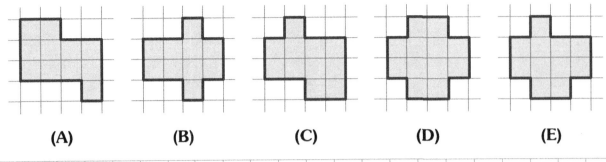

(A) (B) (C) (D) (E)

Problem 3. Masha says, "It's gray. It's a circle or a triangle". What figure is she talking about?

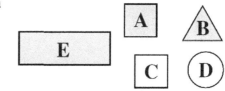

Problem 4. Alexey places two piles of stones on a two-plate scale. What weight should he add to the right plate so that the scale is balanced?

Problem 5. Sasha has been up for an hour and a half. He is happy because he is taking the train in three and a half hours. How many hours before the train left did it get up?

Problem 6. Kesar has 13 coins in his pocket: only 5 cents and 10 cents. Which of the following cannot be the amount Kesar has in his pocket: 70, 80, 115, 125 or 135 cents?

Problem 7. We fold along the thick line to bring the left side to the right side. Which letter will not be covered by a gray box?

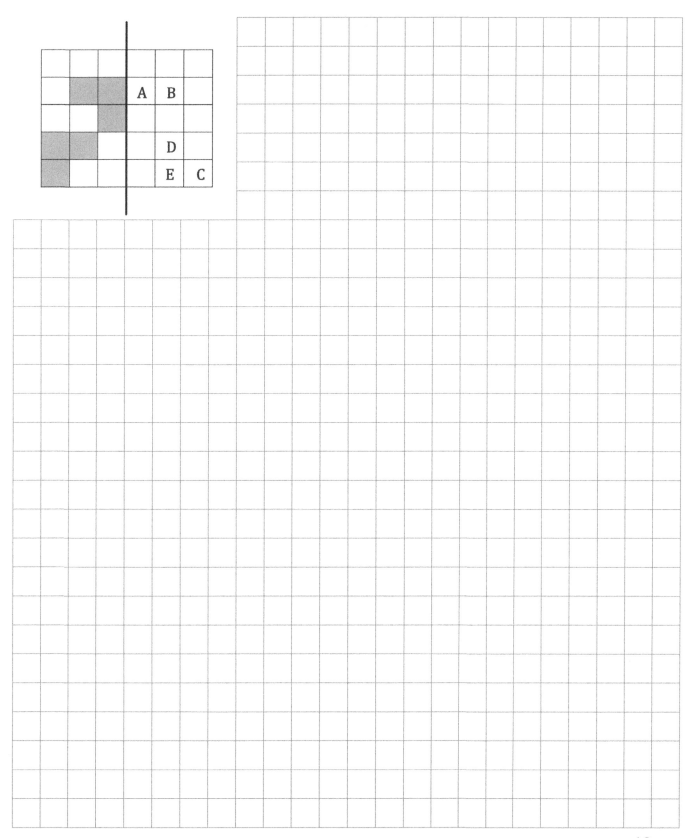

Problem 8. Lesha paid 1 dollar and 50 cents for three croissants. Misha paid 2 dollars and 40 cents for two pies. How much will Vasya pay for a croissant and a pie?

Olympiad 2012

(III Mathematical Olympiad "Unikum")

Problem 1. What is the longest line?

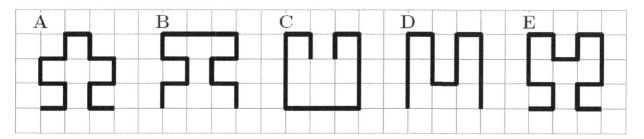

Problem 2. Grisha will hang the clothes on a rope using the fewest possible number of clothespins, as shown in the figure. For example, to hang 3 towels he needs 4 clothespins. How many clothespins will he need to hang 9 towels?

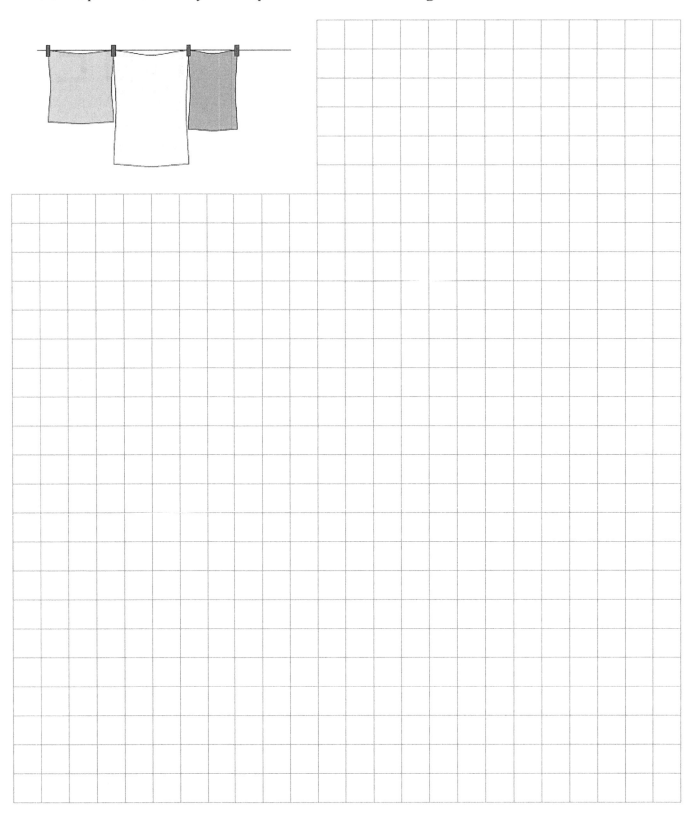

Problem 3. Today, Anna added her age with her sister's age and got 10 as a result. If she does the same operation in a year, what result will she get?

Problem 4. A dragon has three heads. Every time a hero cuts off one of the dragon's heads, three new heads appear. The hero cuts off one head and then cuts off another head. How many heads will the dragon have now?

Problem 5. A game board is made up of a sequence of stars, shamrocks, gifts and trees. Rustam spilled juice on the board making some images disappeared. How many stars were on the board before Rustam spilled the juice?

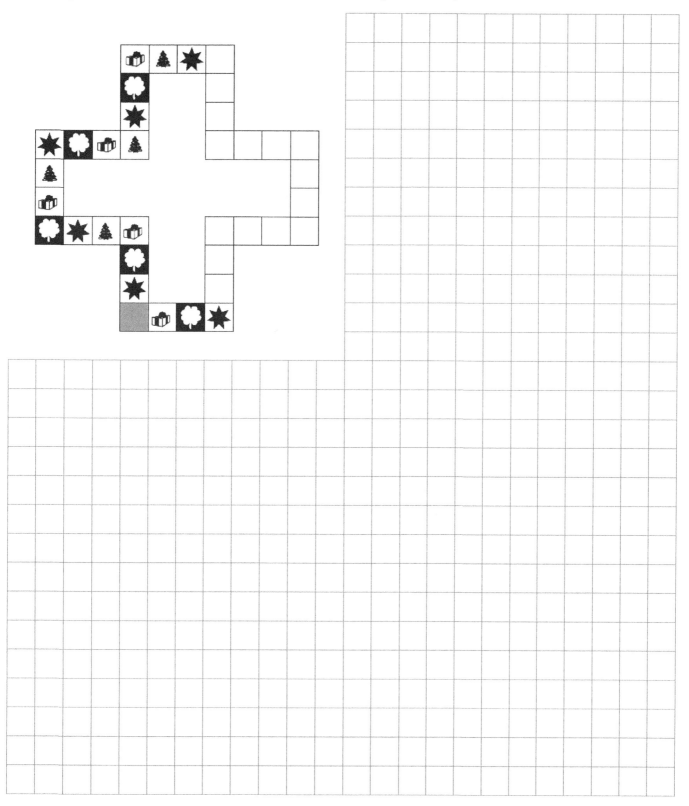

Problem 6. The jumping sparrow likes to play along the fence, jumping from stake to stake. The sparrow does 4 jumps forward, 1 jump backward, again, 4 jumps forward, 1 jumps backward, and so on. Knowing that the sparrow takes 1 second in each of the jumps, how many seconds does it take to go from stake A to stake B?

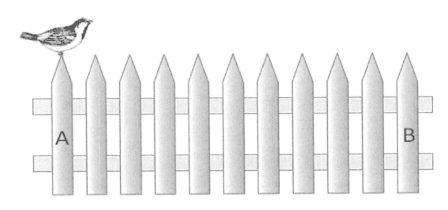

Problem 7. At a school party, Dmitry, Ivan and Boris each received a bag with 10 candies. Each of the boys ate a candy and gave another to the teacher. How many candies did the three boys have in all?

Problem 8. On the board shown below are placed some coins. How many coins must be removed in order to have 2 coins in each row and 2 coins in each column?

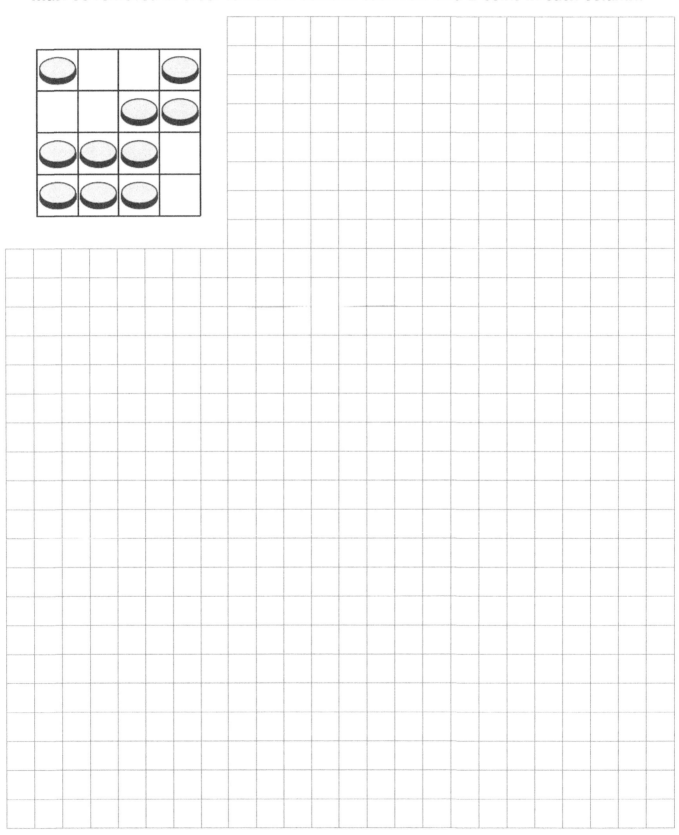

Olympiad 2013

(IV Mathematical Olympiad "Unikum")

Problem 1. Ivan has two kittens of the same weight. What is the weight of a kitten, if Ivan weighs 30 kilograms and with two kittens – 36 kilograms?

Problem 2. A father gave 5 apples to each of his three children (Anna, Sonya and Mikhail). Anna gave Sonya 3 apples, and then Sonya gave half of her apples to Mikhail. How many apples does Mikhail have now?

Problem 3. In a game it is possible to carry out the exchanges presented in the figure. Andrey has 2 pears. How many strawberries will Andrey have after exchanging his two pears according to the suggested rule?

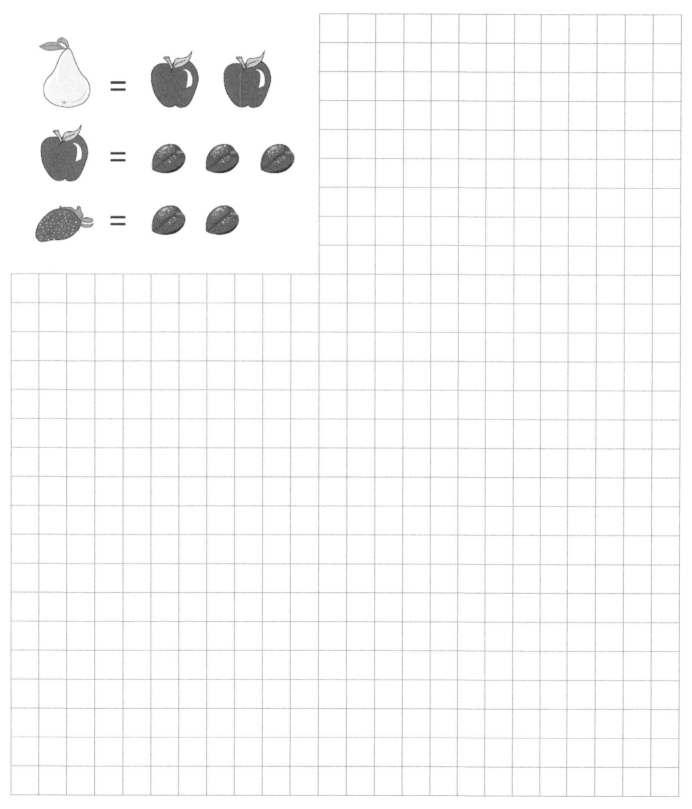

Problem 4. How many carrots can a rabbit that moves freely in accessible areas of the maze eat?

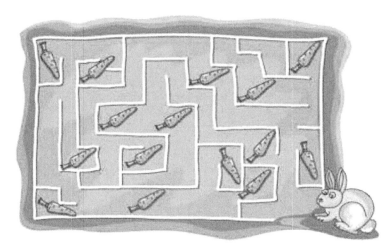

Problem 5. Sofia made a series of 5 houses with matches. In the figure you can see the beginning of this series. How many matches did Sofia use?

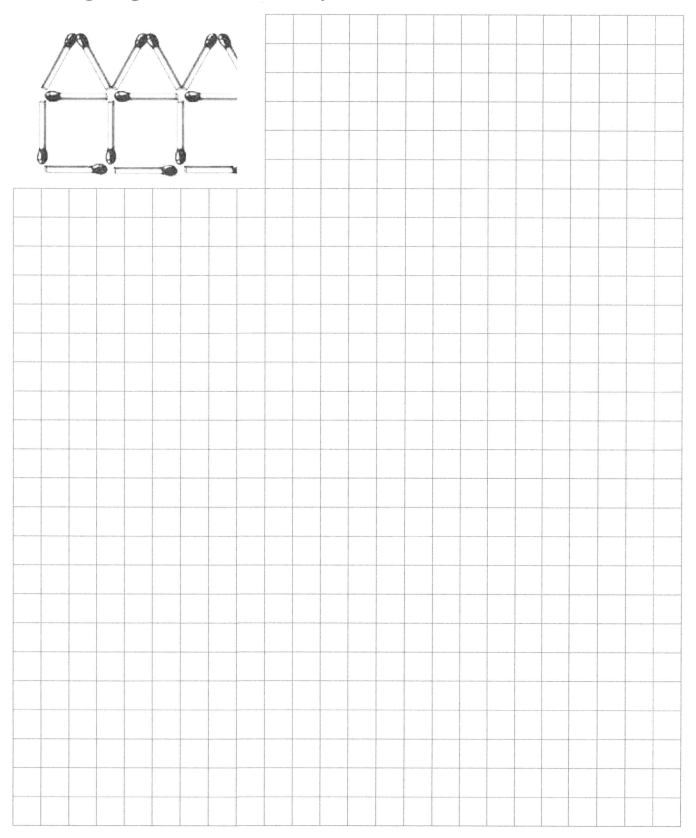

Problem 6. There are five children in a family. Katya is 2 years older than Boris, but 2 years younger than Denis. Tara is 3 years older than Andrey. Boris and Andrey are twins. Which of the children is the oldest?

Problem 7. Mikhail made a large cube out of 27 small cubes and painted it gray. Then, he removed four small cubes as shown in the figure. When the paint has not dried, Mikhail on a sheet of paper made prints of each face of the resulting figure. How many of the proposed prints did he manage to make?

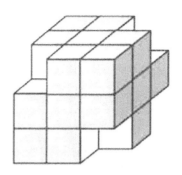

Problem 8. In a box there is a square bar of chocolate, made up of small squares (see figure). Anton ate 20 small squares along the walls of the box. How many small squares of chocolate are left in the box?

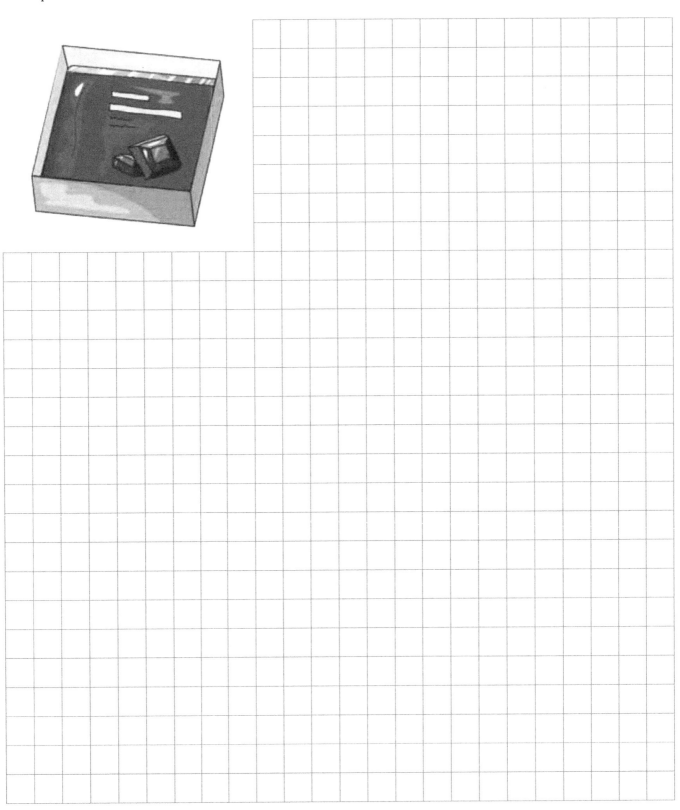

Olympiad 2014

(V Mathematical Olympiad "Unikum")

Problem 1. How many numbers are there between the numbers 10 and 32 that can only be written with the digits 1, 2 or 3? Digits in numbers can be repeated.

Problem 2. Seven strips of paper were placed on the table so that each subsequent one lay above all the previous ones (see figure). The first to be placed on the table was the strip with the number 2. The strip with the number 6 was placed last. Which strip was placed fourth?

Problem 3. How many frogs did the three pelicans catch together?

Problem 4. The chessboard is damaged. How many black squares, to the right of the line, are missing?

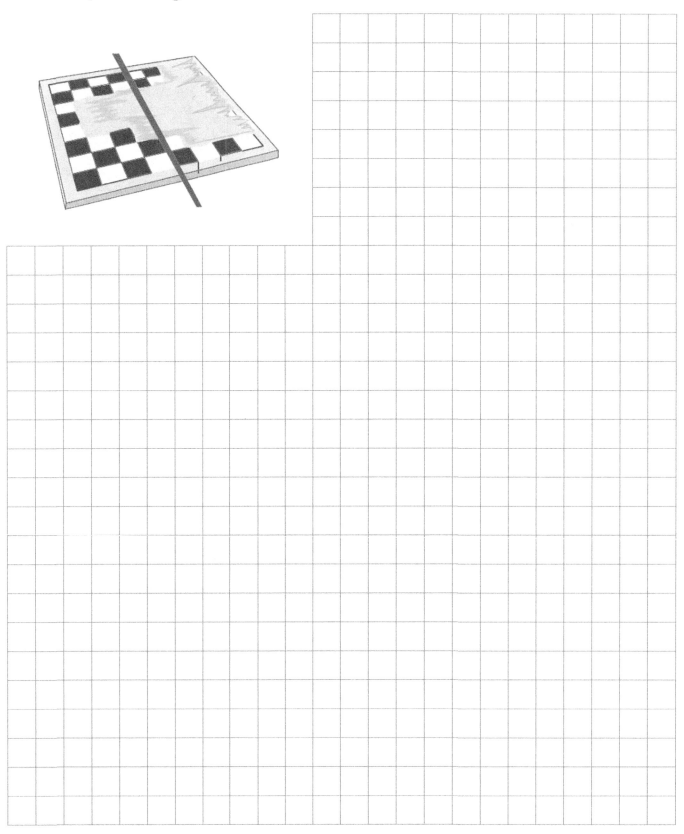

Problem 5. A rabbit eats either 10 carrots or 2 cabbages a day. Last week, the rabbit ate 6 cabbages. How many carrots did it eat last week?

Problem 6. The numbers 2, 3, 4, and 5 must be placed in the cells to get the greatest sum of two two-digit numbers. What is this sum equal to?

Problem 7. With which of the shapes shown in the options, it is not possible to make the shape presented in the figure besides?

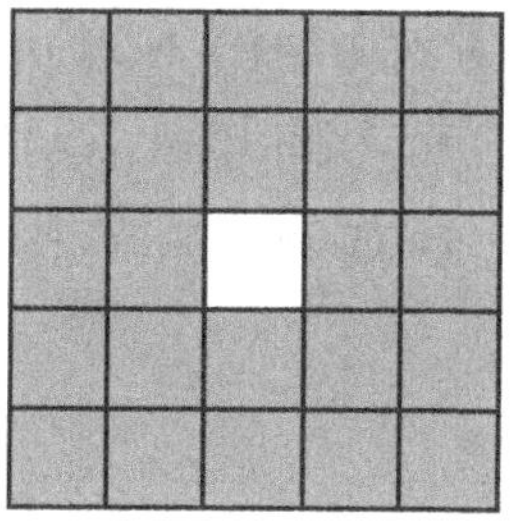

(A)

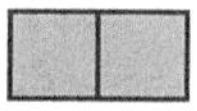

(B)

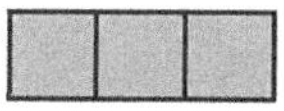

(C)

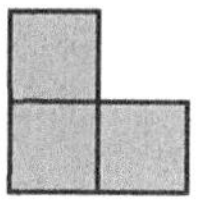

(D)

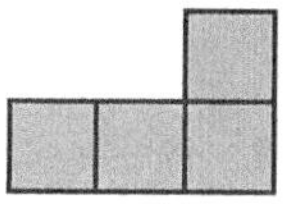

(E)

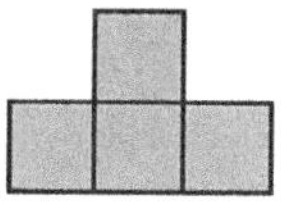

Problem 8. Fyodor has 4 red cubes, 3 green cubes, 2 blue cubes and 1 yellow cube. He builds a tower (see figure) so that two cubes with the same color do not touch each other. What color is the cube with the question mark "?"?

Olympiad 2015

(VI Mathematical Olympiad "Unikum")

Problem 1. Khitrun made two bricks, gluing two identical cubes as shown in the figure opposite. Which of the figures suggested in the options, Khitrun will not be able to make of these two bricks?

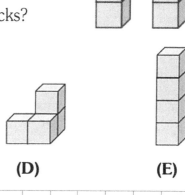

(A)   (B)   (C)   (D)   (E)

Problem 2. Luka has 9 candies, Viktor has 17. How many candies does Viktor have to give to Luka so that both friends have the same number of candies?

Problem 3. Six identical towers were built of gray and white cubes. Each tower is built of five cubes. Cubes of the same color are not attached to each other (see figure). How many white cubes were used to build the towers?

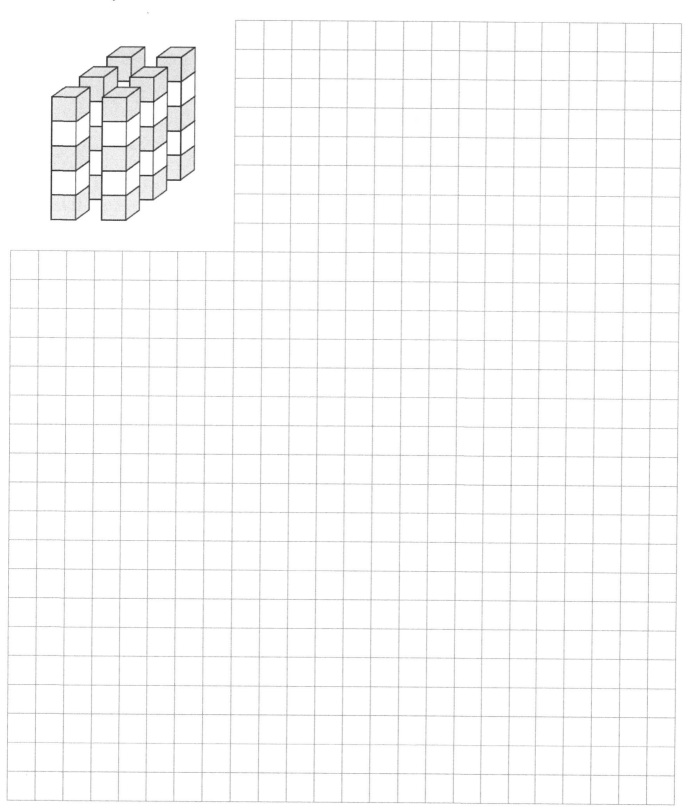

Problem 4. Which tile is missing in the pattern?

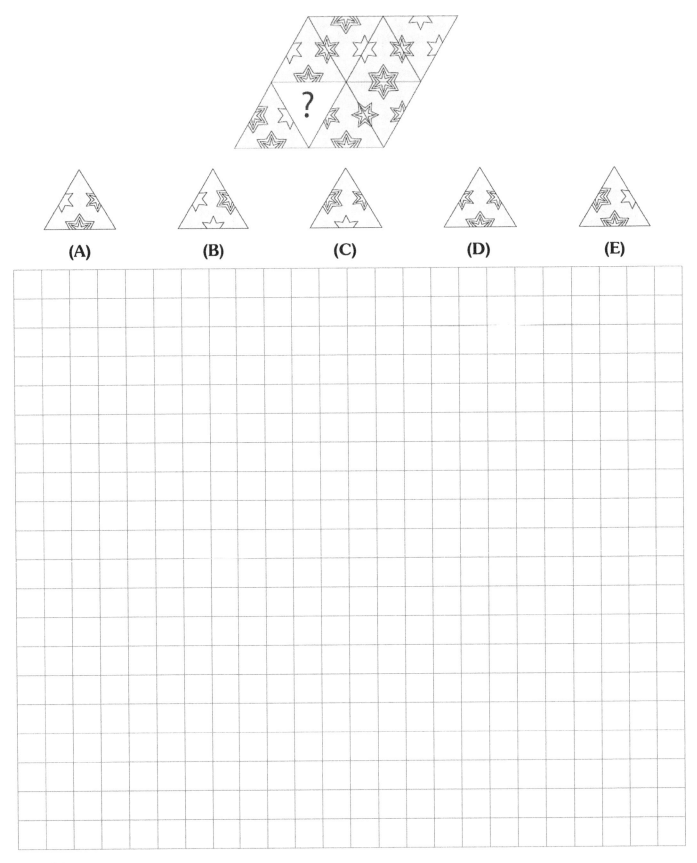

(A)   (B)   (C)   (D)   (E)

Problem 5. The date of May 5, 2015 is recorded as follows: 05.05.2015. Three digits 5 are used in this record. When will three digits 5 be used in the date record for the first time after this date?

Problem 6. Each of the numbers 1, 2, 3, 4, 5 is written one by one in the cells so that all actions are performed correctly. What is the number in the cell with the question mark?

Problem 7. Eleven flags were set up along the straight race track. The first check box was set at the start, and the last – at the finish. The distance between adjacent flags is 8 *m*. What is the length of this route?

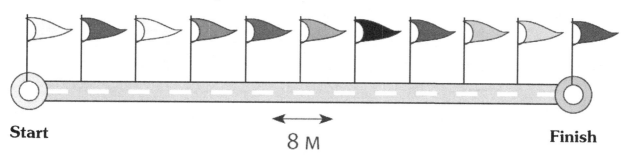

Start   **Finish**

8 M

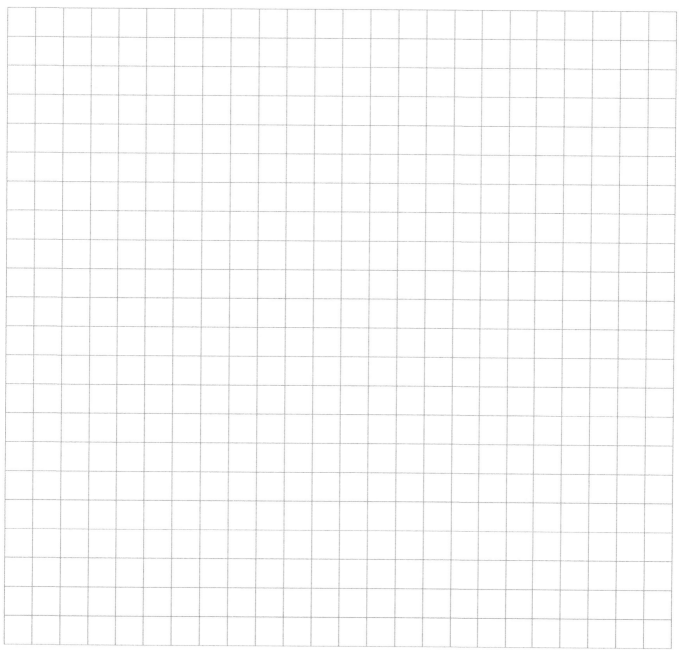

Problem 8. Vasya plays a game where he moves from one circle to the next adjacent circle with one jump. In how many different ways, after four jumps, will Vasya be able to jump from a circle with the letter S (Start) to a circle with the letter F (Finish), if no circle can be visited more than once?

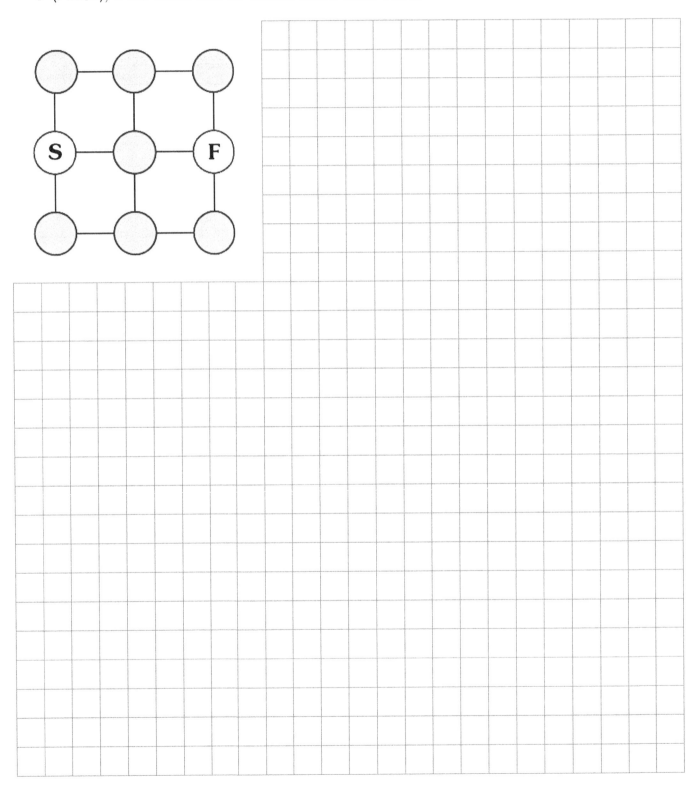

Olympiad 2016

(VII Mathematical Olympiad "Unikum")

Problem 1. The house has twelve rooms. Each room has two windows and one lamp that illuminates the entire room. Eighteen windows were lit up last night. In how many rooms was the light turned off?

Problem 2. Viktor walked along the road and read all the letters located just to his right along the way (see figure below). Going from point 1 to point 2, he read:

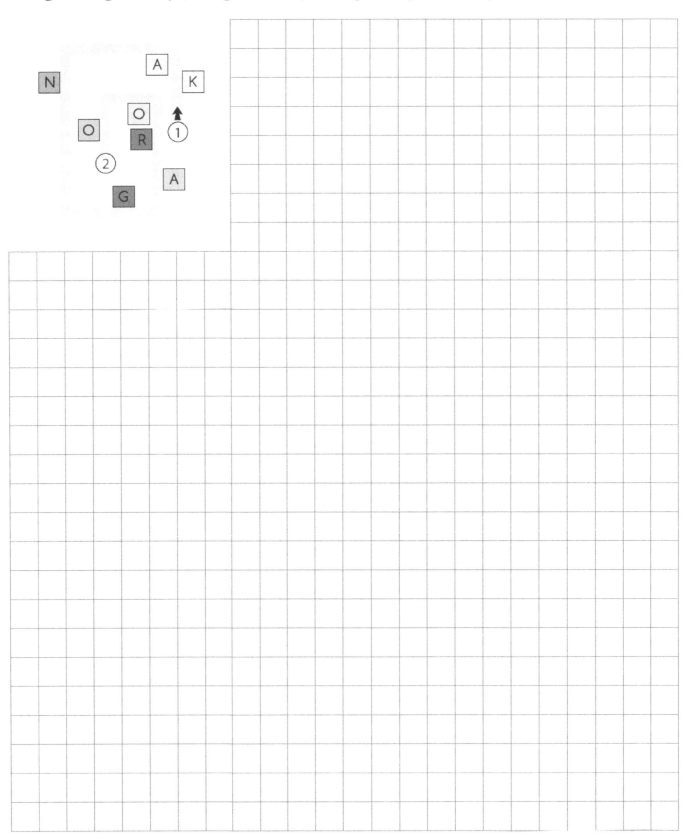

Problem 3. The figure shows four ladybugs  . Each ladybug should perch on a flower according to the rule: the difference between the numbers of dots on its wings is equal to the number of leaves, and the sum of the number of dots on the ladybug's wings is equal to the number of petals. On which flower will not a single ladybug perch?

(A)   (B)   (C)   (D)   (E)

Problem 4. Which of the following options should be inserted in the center of the picture to connect the lines correctly?

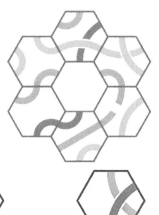

(A)　　　　(B)　　　　(C)　　　　(D)　　　　(E)

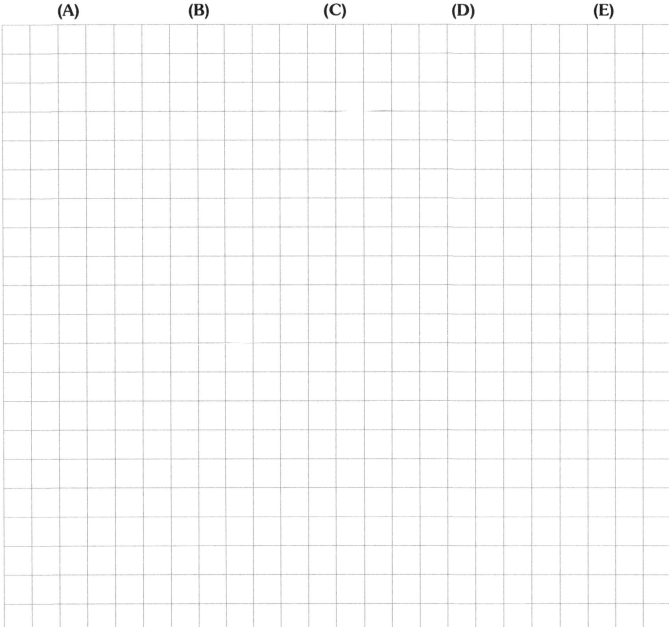

Problem 5. Five sparrows were sitting on a branch (see picture). Some looked to their left, others looked to their right. Each sparrow chirps as many times as it sees sparrows. For example, the fourth sparrow chirped three times. How many times do they chirp in total?

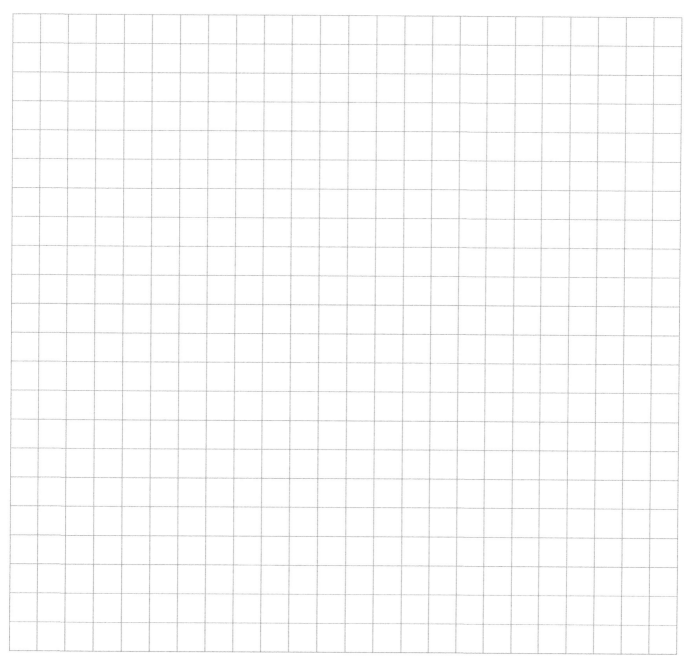

Problem 6. Mudragelyk, Veselun, Lasunchyk, Sonko and Khitrun each live in their own house.

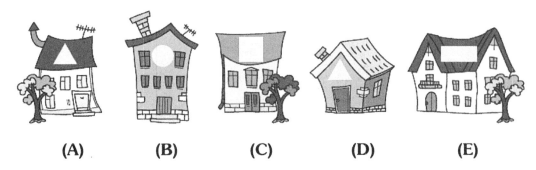

(A)   (B)   (C)   (D)   (E)

Mudragelik and Veselun have only one neighbor. Lasunchyk lives in a house with a triangle on the roof. There is no tree in front of Sonko's house. Veselun lives next to Sonko. In what house does Khitrun live?

Problem 7. Which of the following pictures can we make with all five of the cards shown below? Cards can be put on top of each other, but cannot be cut.

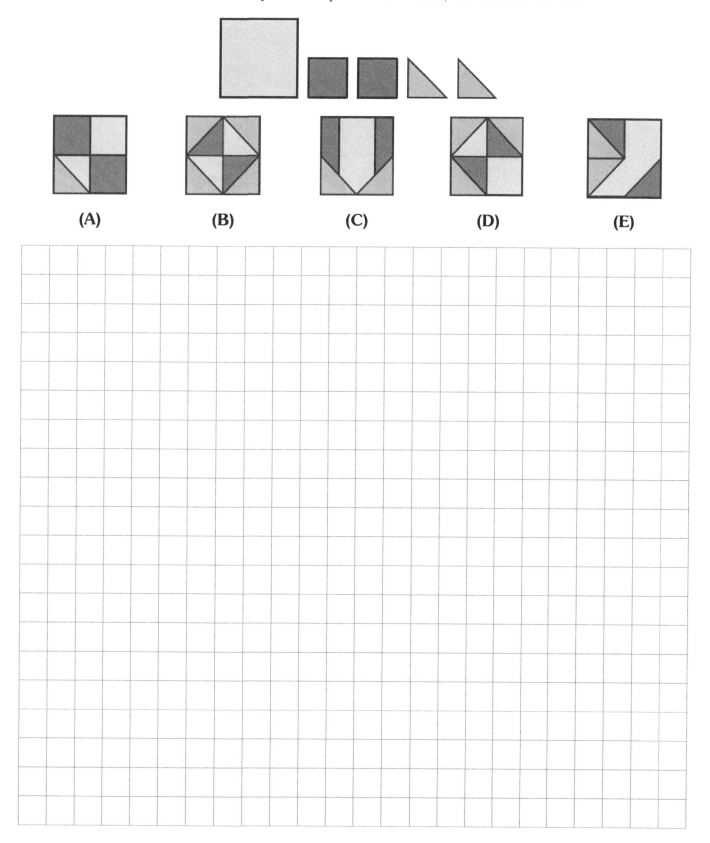

Problem 8. What is the largest number of groups into which the numbers 1, 5, 8, 9, 10, 12, and 15 can be divided, so that in all groups the sums of the numbers are equal?

Olympiad 2017

(VIII Mathematical Olympiad "Unikum")

Problem 1. Alexey was turning the shape on a table in the same direction, as shown in the figure. What is the position of the shape after the sixth turn?

Problem 2. Taras and Vasya line up in the school canteen. Taras knows that there are exactly 7 students in front of him. Taras is in front of Vasya. Vasya knows that a total of 11 students line up with him. How many students are behind Vasya?

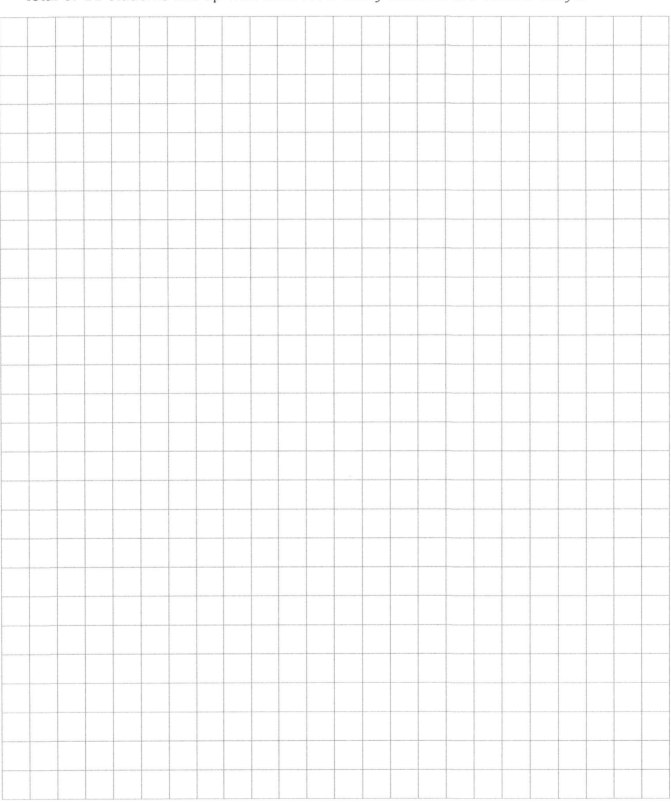

Problem 3. Igor wants to make three copies of a crown using stickers ⬭, ✚ and ▭. If the store sells sheets with stickers and . What is the least number of sheets with stickers Igor needs to buy?

Problem 4. Ivan has two identical shapes, as shown in the figure on the right. Which of the proposed options could represent the shape made with the previous ones?

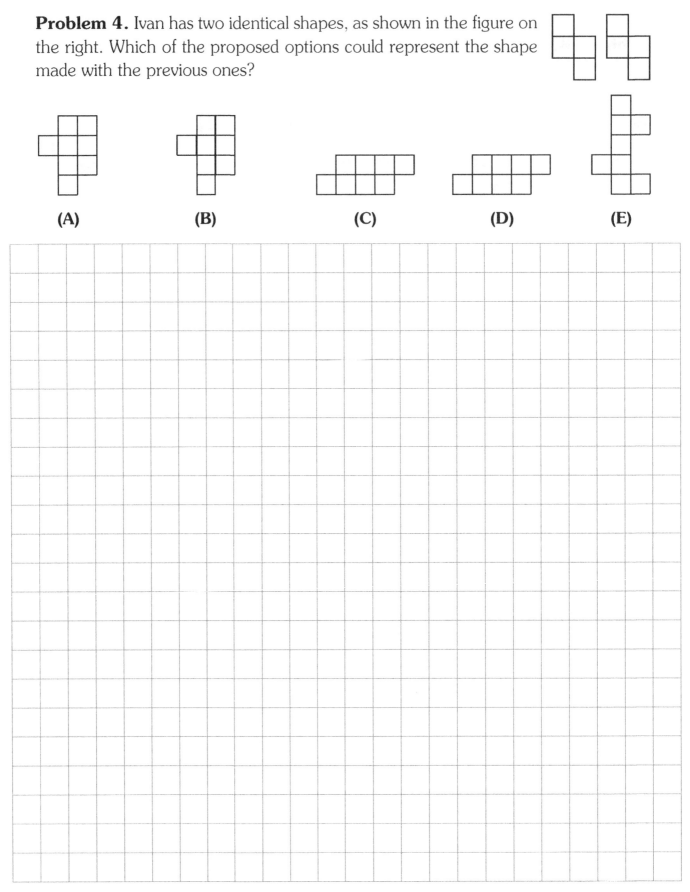

(A)   (B)   (C)   (D)   (E)

Problem 5. A kangaroo performs 10 jumps in 1 minute and then rests for 3 minutes. Then, it performs 10 jumps in 1 minute again and then rest for 3 minutes and so on. What is the least amount of minutes the kangaroo needs to do 30 jumps?

Problem 6. What stamp should be used to obtain on a sheet the picture shown in the figure besides?

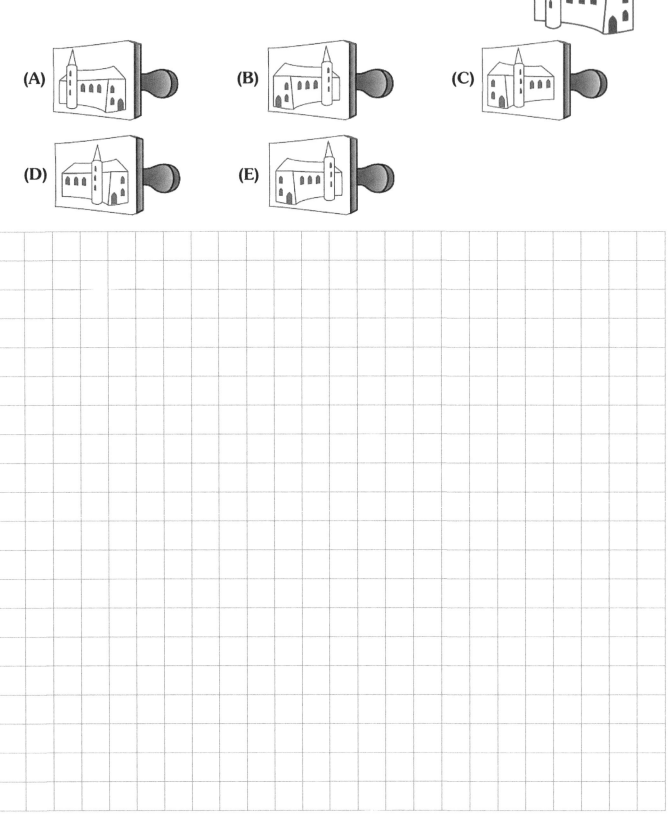

Problem 7. Each of the four keys opens only one of the four locks. The digits on the keys correspond to the inscribed letters of the lock to be unlocked. In addition, different digits correspond to different letters. What is the inscription on the last lock?

Problem 8. Lesha put six of her toys on a shelf with six compartments, putting a single toy in each compartment. It is known that: The bear is between the doll and the rabbit in the same row, the doll is on top of the car, the kangaroo to the left of the ball and to the right of the car in the same row. What toy will be in the compartment with the question mark?

Olympiad 2018

(IX Mathematical Olympiad "Unikum")

Problem 1. Petya drew a pattern twice, as you can see in the figure below. If he draws the same pattern again, what point will be on his drawing?

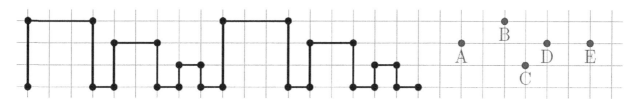

Problem 2. Lesha has the four pieces represented in the figure below to complete the puzzle, but she will only need 3. Which piece will not be necessary?

A   B   C   D

Problem 3. The two sketches in transparent sheet shown in the figure besides are placed one above the other. Which of the following figures can be obtained?

(A)   (B)   (C)   (D)   (E)

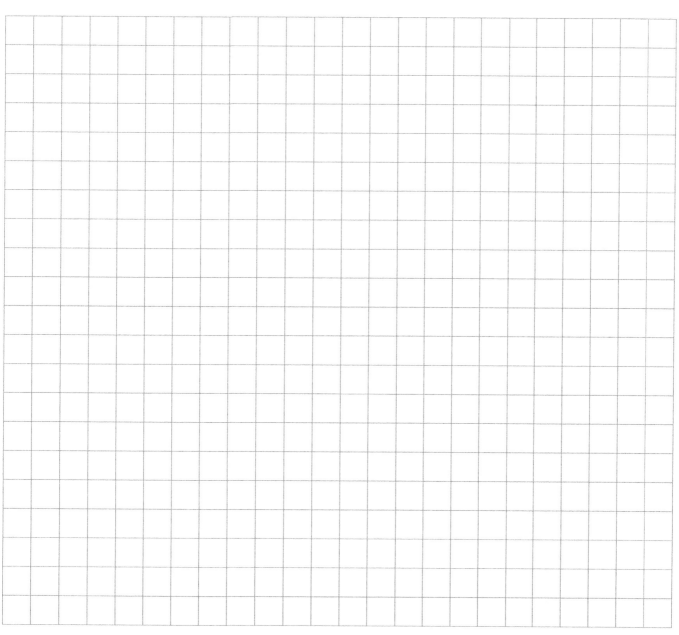

Problem 4. Irina built some towers following the pattern indicated in the figure opposite. What was the sixteenth tower to be built?

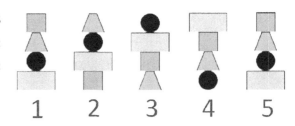

(A)  **(B)**  **(C)**  **(D)**  **(E)**

Problem 5. Dasha was shooting arrows at a target. In the first shoot she got 6 points with the 3 arrows placed on the target, as shown in the figure on the left. On the second throw she got 8 points as shown in the center figure. If the figure on the right represents the result of her third shoot, how many points did she get this time?

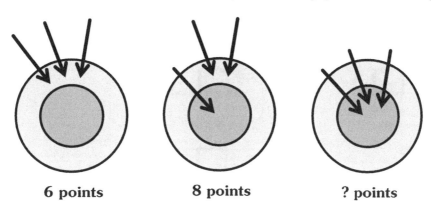

6 points   8 points   ? points

Problem 6. The dog represented in the figures below wants to go and eat the bone that is waiting for it, but for that it has to follow one of the paths indicated in the figures. Knowing that at intersections it will have to turn exactly 3 times to the right and 2 times to the left, which path will it have to choose?

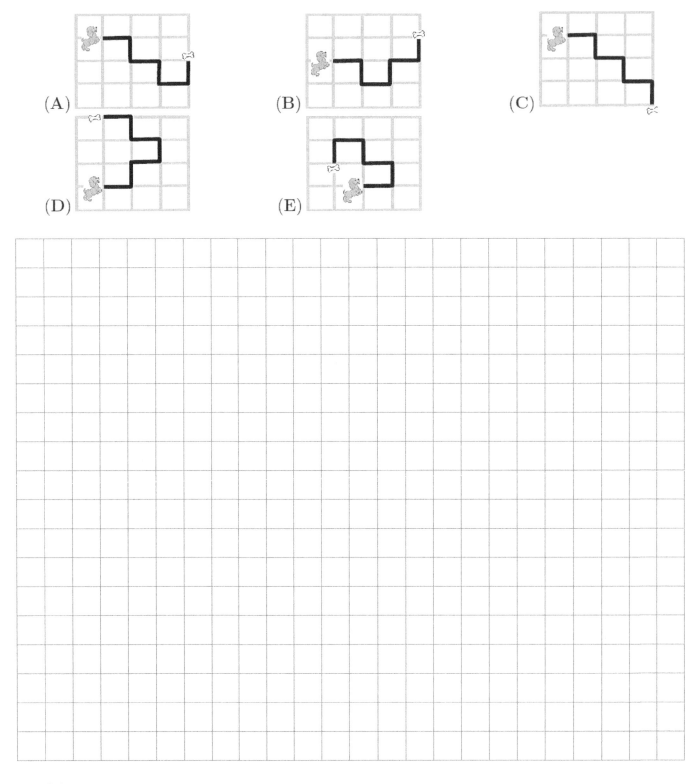

Problem 7. In the enchanted garden of Mathematics, the number of goblins that can shelter under a mushroom is equal to the number of spots that the mushroom's hat has. In the figure below we see one side of the mushroom hats in this garden and it is known that the other side of the hat has exactly the same number of spots. If there are 30 goblins in the garden on a rainy day, how many goblins will not be able to shelter under the mushrooms?

Problem 8. At the ice-cream parlor "Frozen Delight" an ice cream costs 1 ruble, but now there is a promotion so that six ice creams cost 5 rubles. What is the largest number of ice creams you can buy for 36 rubles?

Olympiad 2019

(X Mathematical Olympiad "Unikum")

Problem 1. The four strips are glued to the clear glass as shown in the figure. A hole is made in the upper left corner of the glass. Which picture shows what will be seen from the back?

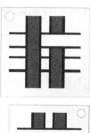

(A)   (B)   (C)   (D)   (E)

Problem 2. Each of the shapes depicted in the options is made by gluing four cubes of the same size. It is necessary to paint all the faces of the shapes. Andrey spends the same amount of paint on each face of the cube. Which shape will Andrey spend the least amount of paint on?

(A)   (B)   (C)   (D)   (E)

Problem 3. The floor is covered with the same rectangular tiles as shown in the figure. The smaller side of each tile is $1\,m$. What is the length of the floor?

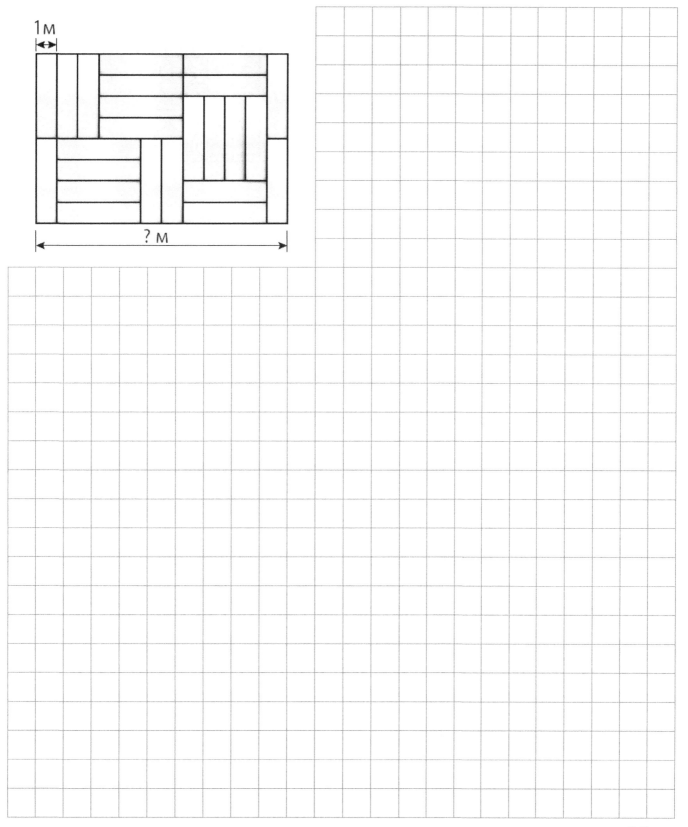

Problem 4. Boris covers the square depicted on the right with the shapes depicted on the left. The numbers in the shapes superimpose the same numbers in the square. What shape is missing?

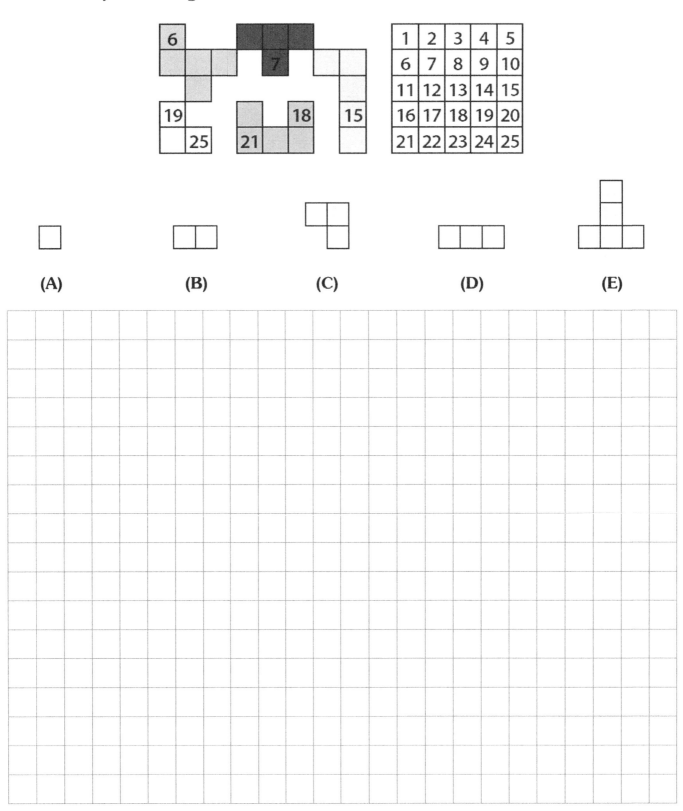

(A)　　　(B)　　　(C)　　　(D)　　　(E)

Problem 5. The difference between the ages of Andrey and his older sister Elena is 7 years. Andrey is 8 years younger than Masha. Place the three children in descending order of age.

Problem 6. The figure shows nine squares:

First, Natasha painted all the black squares in white. Then Boris painted all the gray squares in black, and finally Mikhail painted all the white squares in gray. What picture did they get in the end?

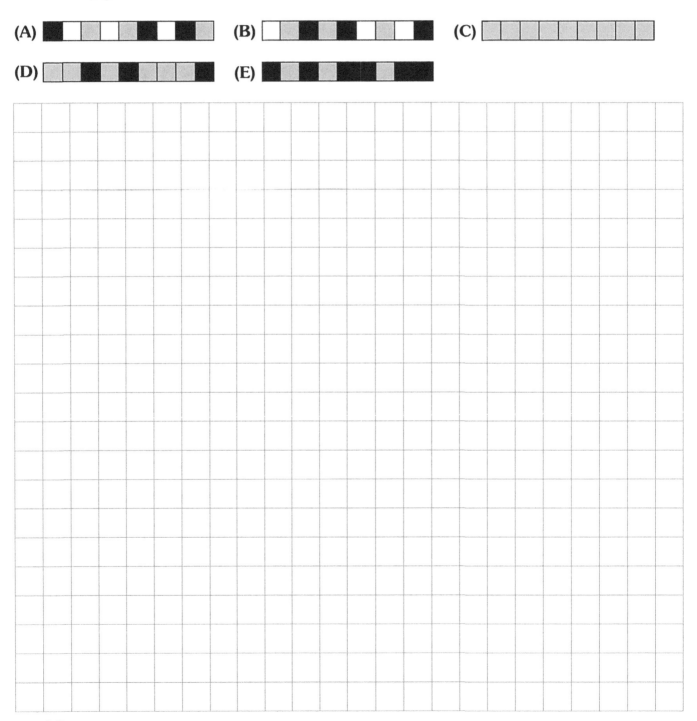

Problem 7. In the square shown in the figure below, Petya chooses a square of size 2×2 so that the sum of the four numbers in this square is greater than 49. What numbers should be in the selected square?

1	2	3	4
5	6	7	8
9	10	11	12
13	14	15	16

Problem 8. The magic ball converter transforms one red ball into three white or one white into two red. Galenka has three red balls and one white. If she uses the converter three times. What is the smallest number of balls she can get?

Olympiad 2020

(XI Mathematical Olympiad "Unikum")

Problem 1. Masha wrote six numbers 1, 2, 3, 4, 5 and 6 in six different cells as shown in the figure on the right. The sum of the numbers in the dark grey cells is 10, the sum of the numbers in the light grey cells is also 10. What number did the girl write in the question mark cell?

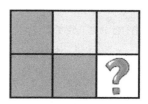

Problem 2. Two identical trains, each of which has 31 cars, go to meet each other and arrive at the station on adjacent tracks. If car № 19 of the first train is in front of car №19 of the second train, then which car will be in front of car № 12?

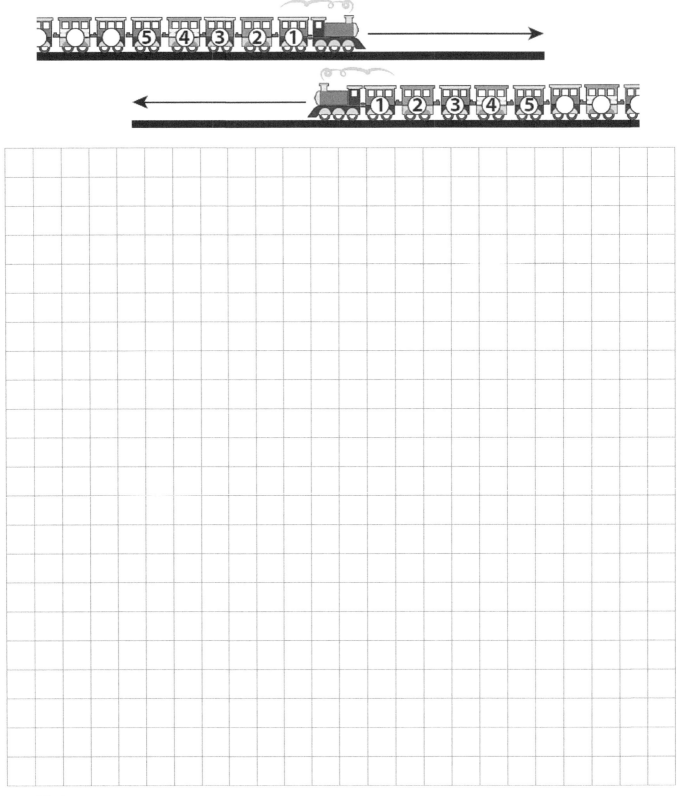

Problem 3. Mikhail tries to cut the depicted figure into as many "corners" as possible

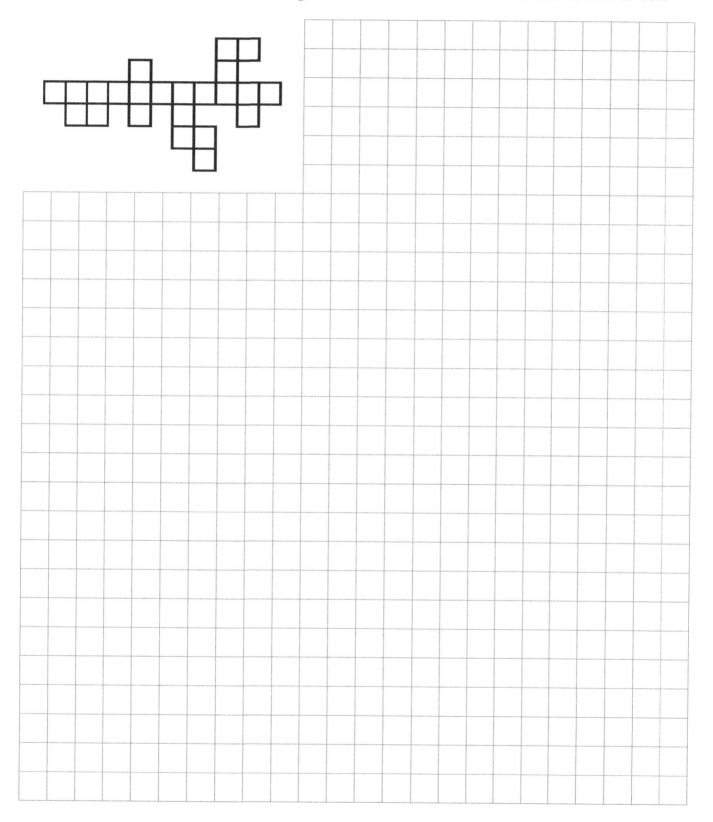

. What is the largest number of such corners he will be able to cut?

Problem 4. Ruslan has 9 cards 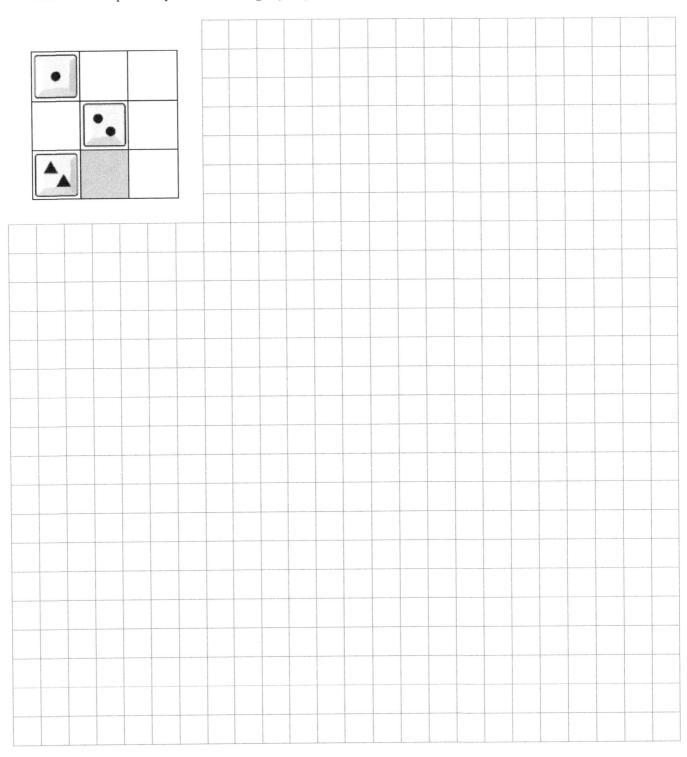. He places them on the board so that any two cards in each row and in each column differ in both the geometric shape depicted on them and their number. What card will Ruslan put in place of the gray square?

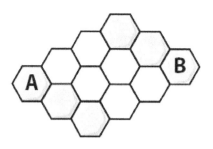

Problem 5. Maya the bee can fly from the grey hexagonal honeycomb only to the adjacent grey one with a common side. In how many different ways can only two white honeycombs be painted grey so that Maya the Bee can get from honeycomb A to honeycomb B?

Problem 6. The arrows in the figure point from one kangaroo to another and indicate that the first kangaroo is taller than the second. In particular, "kangaroo F" is taller than "kangaroo C". Which kangaroo is the shortest?

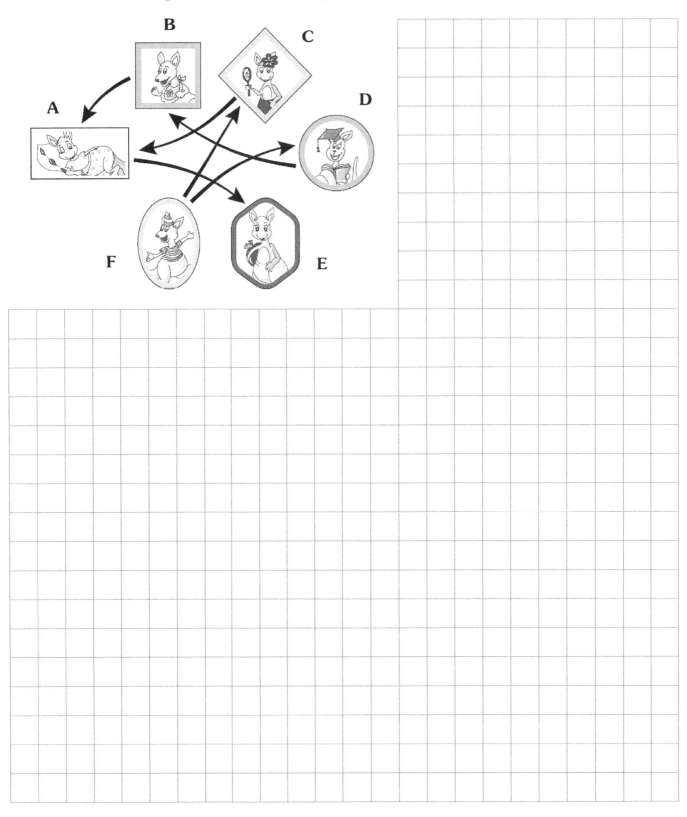

Problem 7. There are a few apples and 8 pears in the basket. They are all either yellow or green. Apples are 3 more than green fruits. There are 6 yellow pears. How many yellow apples are in the basket?

Problem 8. Vasya and Tatiana exchanged candies. At first Vasya gave Tatiana as many candies as she had. After that, Tatiana gave Vasya as many candies as he had after the first exchange. After these exchanges, each of them has 4 candies. How many candies did Vasya have in the beginning?

Answers

Olympiad 2011

1. 2110.

2. Option (D).

3. Figure B.

4. 9 *kg*.

5. 5 hours.

6. 135 cents.

7. Letter E.

8. 1 dollar 70 cents.

Olympiad 2012

1. Option (E).

2. 10 clothespins.

3. 12 years.

4. 7 heads.

5. 9 stars.

6. 14 secons.

7. 24 candies.

8. 2 coins.

Olympiad 2013

1. 3 *kg.*

2. 9 apples.

3. 6 strawberries.

4. 8 carrots.

5. 26 matches.

6. Denis.

7. 4 prints.

8. 16 small squares.

Olympiad 2014

1. 7 numbers.

2. Strip 3.

3. 9 frogs.

4. 12 black squares.

5. 40 carrots.

6. 95.

7. Option (E).

8. Red.

Olympiad 2015

1. Option (B).

2. 4 candies.

3. 12 white cubes.

4. Option (C).

5. May 15, 2015.

6. The number is 5.

7. 80 *m*.

8. 6 different ways.

Olympiad 2016

1. 3 rooms.

2. KNAO.

3. Option (E).

4. Option (A).

5. 10 times.

6. Option (C).

7. Option (A).

8. 3 groups.

Answers

Olympiad 2017

1. The figure bellows shows the position.

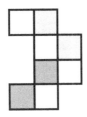

2. 2 students.

3. 4 sheets.

4. Option (C).

5. 9 minutes.

6. Option (E).

7. GAG.

8. The rabbit.

Olympiad 2018

1. Point D.

2. Piece A.

3. Option (A).

4. Option (E).

5. 12 points.

6. Option (C).

7. 2 goblins.

8. 43 ice creams.

Olympiad 2019

1. Option (B).

2. Option (B).

3. 12 *m*.

4. Option (C).

5. Masha, Elena, Andrey.

6. Option (D).

7. 10-11-14-15 or 11-12-15-16.

8. 8 balls.

Olympiad 2020

1. The number is 1.

2. The car № 26.

3. 7 corners.

4. The card is .

5. 5 different ways.

6. Kangaroo E.

7. 5 yellow apples.

8. 5 candies.

Answers

Made in the USA
Las Vegas, NV
03 February 2024

85247804R00070